There are about
200 kinds.
They look different!

My Favorite Animal

I LOVE MONKEYS!

By Beth Gottlieb

Gareth Stevens Publishing

Please visit our website, www.garethstevens.com. For a free color catalog of all our high-quality books, call toll free 1-800-542-2595 or fax 1-877-542-2596.

Library of Congress Cataloging-in-Publication Data
Names: Gottlieb, Beth, author.
Title: I love monkeys! / Beth Gottlieb.
Description: Buffalo, New York : Gareth Stevens Publishing, [2023] | Series: My favorite animal | Includes index.
Identifiers: LCCN 2022014046 (print) | LCCN 2022014047 (ebook) | ISBN 9781538283318 (paperback) | ISBN 9781538283332 (library binding) | ISBN 9781538283349 (ebook)
Subjects: LCSH: Monkeys–Juvenile literature.
Classification: LCC QL737.P925 G68 2023 (print) | LCC QL737.P925 (ebook) | DDC 599.8–dc23/eng/20220516
LC record available at https://lccn.loc.gov/2022014046
LC ebook record available at https://lccn.loc.gov/2022014047

First Edition

Published in 2023 by
Gareth Stevens Publishing
2544 Clinton St,
Buffalo, NY 14224

Copyright © 2023 Gareth Stevens Publishing

Editor: Kristen Nelson
Designer: Rachel Rising

Photo credits: Cover, p. 1 ScottWalmsley/Shutterstock.com; p. 5 Xenia_Photography/Shutterstock.com; p. 7 KAMONRAT/Shutterstock.com; p. 7 anut21ng Stock/Shutterstock.com; p. 7 Yusnizam Yusof/Shutterstock.com; p. 7 Ondrej Prosicky/Shutterstock.com; p. 9 Carlos Ezequiel Vannoni/Shutterstock.com; p. 11 Dave Primov/Shutterstock.com; pp. 13, 24 VittoriaChe/Shutterstock.com; pp. 15, 24 MirasWonderland/Shutterstock.com; p. 17 Laura Kathleen Lewis/Shutterstock.com; p. 19 Curioso.Photography/Shutterstock.com; p. 21 apiguide/Shutterstock.com; p. 23 Smolina Marianna/Shutterstock.com.

All rights reserved. No part of this book may be reproduced in any form without permission in writing from the publisher, except by a reviewer.

Printed in the United States of America

CPSIA compliance information: Batch #CWGS23: For further information contact Gareth Stevens, New York, New York at 1-800-542-2595.

Find us on

Contents

Monkeys are my favorite animal.

They live all over
the world.
Most live in trees.

Baboons live on the ground.

Monkeys live in groups.
These are troops.

Monkeys have hair.
They have tails.

Spider monkeys have
a long tail.
They use it to hold on
to tree branches.

Monkeys can be
big or small.
Mandrills are
the biggest.

Monkeys eat plants.
They eat fruits
and leaves.
Some eat bugs!

I love monkeys!

Words to Know

tail

troop

Index